MEASURING
INTERNATIONAL
TRADE
ON U.S. HIGHWAYS

Panel on Bureau of Transportation Statistics
International Trade Traffic

Joel L. Horowitz and Tom Plewes, Editors

Committee on National Statistics

Division of Behavioral and Social Sciences and Education

NATIONAL RESEARCH COUNCIL
OF THE NATIONAL ACADEMIES

THE NATIONAL ACADEMIES PRESS
Washington, D.C.
www.nap.edu

THE NATIONAL ACADEMIES PRESS 500 Fifth Street, N.W. Washington, DC 20001

NOTICE: The project that is the subject of this report was approved by the Governing Board of the National Research Council, whose members are drawn from the councils of the National Academy of Sciences, the National Academy of Engineering, and the Institute of Medicine. The members of the committee responsible for the report were chosen for their special competences and with regard for appropriate balance.

This study was supported by Contract/Grant No. DTTS59-01-C-00402 between the National Academy of Sciences and the U.S. Department of Transportation. Support of the work of the Committee on National Statistics is provided by a consortium of federal agencies through a grant from the National Science Foundation (Number SBR-0112521). Any opinions, findings, conclusions, or recommendations expressed in this publication are those of the author(s) and do not necessarily reflect the views of the organizations or agencies that provided support for the project.

International Standard Book Number 0-309-09519-0 (Book)
International Standard Book Number 0-309-54706-7 (PDF)

Additional copies of this report are available from the National Academies Press, 500 Fifth Street, N.W., Lockbox 285, Washington, DC 20055; (800) 624-6242 or (202) 334-3313 (in the Washington metropolitan area); Internet, http://www.nap.edu.

Printed in the United States of America.

Suggested citation: National Research Council. (2005). *Measuring International Trade on U.S. Highways.* Panel on Bureau of Transportation Statistics International Trade Traffic. J.L. Horowitz and T. Plewes, Editors. Committee on National Statistics. Division of Behavioral and Social Sciences and Education. Washington, DC: The National Academies Press.

THE NATIONAL ACADEMIES
Advisers to the Nation on Science, Engineering, and Medicine

The **National Academy of Sciences** is a private, nonprofit, self-perpetuating society of distinguished scholars engaged in scientific and engineering research, dedicated to the furtherance of science and technology and to their use for the general welfare. Upon the authority of the charter granted to it by the Congress in 1863, the Academy has a mandate that requires it to advise the federal government on scientific and technical matters. Dr. Bruce M. Alberts is president of the National Academy of Sciences.

The **National Academy of Engineering** was established in 1964, under the charter of the National Academy of Sciences, as a parallel organization of outstanding engineers. It is autonomous in its administration and in the selection of its members, sharing with the National Academy of Sciences the responsibility for advising the federal government. The National Academy of Engineering also sponsors engineering programs aimed at meeting national needs, encourages education and research, and recognizes the superior achievements of engineers. Dr. Wm. A. Wulf is president of the National Academy of Engineering.

The **Institute of Medicine** was established in 1970 by the National Academy of Sciences to secure the services of eminent members of appropriate professions in the examination of policy matters pertaining to the health of the public. The Institute acts under the responsibility given to the National Academy of Sciences by its congressional charter to be an adviser to the federal government and, upon its own initiative, to identify issues of medical care, research, and education. Dr. Harvey V. Fineberg is president of the Institute of Medicine.

The **National Research Council** was organized by the National Academy of Sciences in 1916 to associate the broad community of science and technology with the Academy's purposes of furthering knowledge and advising the federal government. Functioning in accordance with general policies determined by the Academy, the Council has become the principal operating agency of both the National Academy of Sciences and the National Academy of Engineering in providing services to the government, the public, and the scientific and engineering communities. The Council is administered jointly by both Academies and the Institute of Medicine. Dr. Bruce M. Alberts and Dr. Wm. A. Wulf are chair and vice chair, respectively, of the National Research Council.

www.national-academies.org

Acknowledgments

The Committee on National Statistics (CNSTAT) has a long-standing record of assisting federal statistical agencies in problems at the intersection of statistical methods and public policy. Recognizing this, Congress mandated that the Bureau of Transportation Statistics (BTS), as part of its evaluation of the development of estimates of the ton-miles and value-miles of international freight transported by highways in the United States, support a formal review of its work by the committee. Consequently, in 2003, CNSTAT organized a panel study to evaluate the estimates and the BTS evaluation of them (Bureau of Transportation Statistics, 2004).

The panel included experts in statistical methods, data used in support of fund allocation formula programs, transportation data, emerging technologies for transportation data collection, and survey methodology. The panel met four times, including a workshop that was held November 20-21, 2003. The panel issued an interim report in February 2004, and this is its final report.

The panel would like to thank the BTS staff for their support. In particular, we would like to thank Rick Kowalewski, acting BTS director, and Bill Bannister and Irwin Silberman for their valuable input. Ho-Ling Hwang and Paul Metaxatos presented valuable information at the panel's first meeting in September of 2003. In November 2003, the panel held a workshop. At that workshop, the following people, including BTS staff and others, provided presentations that were tremendously important to our work: Irwin Silberman, Frank Southworth, Antonio Esteve, Catherine Lawson, M.J. Fiocco, Paul Ciannavei, Rob Tardif, Rolf Schmidt, John Fowler, Bill Davie, Harvey Monk, and Dan Melnick. Also, Richard Easley provided a report on emerging technologies for measuring the weight, value, and distance traveled for freight carried by highway that the panel utilized extensively in its section of the report on this topic.

Thanks are also owed to CNSTAT staff members who supported the work of the panel. Tom Plewes served as the study director, bringing his outstanding management skills to the task of organizing informative meetings, assembling relevant background materials to inform the panel, and producing the panel's interim report. Allison Shoup, senior program assistant, provided excellent logistical support to the panel. Michael Cohen, senior program officer, provided invaluable technical assistance in drafting the final report and guiding it through the National Academies review process. Eugenia Grohman, associate director for reports of the Division of Behavioral and Social Sciences and Education, ably edited the report.

Finally, it was a pleasure to work with the panel members, who individually and as a group brought substantial expertise and wisdom to the task.

This report has been reviewed in draft form by individuals chosen for their diverse perspectives and technical expertise, in accordance with procedures approved by the National Research Council's Report Review Committee. The purpose of this independent review is to provide candid and critical comments that will assist the institution in making its published report as sound as possible and to ensure that the report meets institutional standards for objectivity, evidence, and responsiveness to the study charge. The review comments and draft manuscript remain confidential to protect the integrity of the deliberative process. We thank the following individuals for their review of this report: Kenneth D. Boyer, Department of Economics, Michigan State University; James R. Chromy, Statistics Research Division, RTI International, Research Triangle Park, NC; Jose Holguin-Veras, Department of Civil and Environmental Engineering, Rensselaer Polytechnic Institute, Troy, NY; Steven R. Kale, Economics, Oregon Department of Transportation, Salem, OR; Catherine T. Lawson, Geography and Planning Department, University at Albany, Albany, NY; Daniel McCaffrey, Education Group, RAND, Pittsburgh, PA; Michael D. Meyer, School of Civil and Environmental Engineering, Georgia Institute of Technology,

Atlanta, GA; and Alan M. Zaslavsky, Department of Health Care Policy, Harvard Medical School, Boston, MA.

Although the reviewers listed above have provided many constructive comments and suggestions, they were not asked to endorse the conclusions or recommendations nor did they see the final draft of the report before its release. The review of this report was overseen by Robert Groves, Survey Research Center, University of Michigan. Appointed by the National Research Council, he was responsible for making certain that an independent examination of this report was carried out in accordance with institutional procedures and that all review comments were carefully considered. Responsibility for the final content of this report rests entirely with the authoring committee and the institution.

Joel L. Horowitz, *Chair*
Panel on BTS International Trade Traffic

Contents

Executive Summary

In 2003 Congress directed the Bureau of Transportation Statistics (BTS) to carry out a study: "(1) to measure the ton-miles and value-miles of international trade traffic carried by highway for each state; (2) to evaluate the accuracy and reliability of such measures for use in the formula for highway apportionments . . . ; (3) to evaluate the accuracy and reliability of the use of diesel fuel data as a measure of international trade traffic by state; and (4) to identify needed improvements in long-term data collection programs. . . ." The legislation also directed BTS to request that the Committee on National Statistics convene a panel study to review the findings and recommendations of the BTS study and to explore additional data sources and methods for providing improved estimates for use in apportionment formulas. A key purpose of this work is to find appropriate formulas for allocating highway funds to states on the basis of international trade traffic.

There are no direct measures of ton-miles and value-miles of international trade traffic carried by highway in each state. Therefore, BTS had to use models to produce estimates, using as input data from a multitude of public and private sources, none of which were specifically designed for the task. In addition, as directed by the legislation, BTS examined the possibility of using diesel fuel data to directly predict international trade traffic by highway.

The BTS report concluded that its estimates should not be used as input into fund allocation formula programs because of the poor quality of the estimates. To support its conclusion, BTS identified a number of data deficiencies and the need to rely on strong assumptions in the models. BTS also concluded that the relationship between diesel fuel data and international trade traffic carried by highway was not sufficiently strong to adequately differentiate between states in modeling ton-miles or value-miles of international trade traffic. The BTS report included some suggestions on how to address the various data deficiencies, including the introduction of new surveys and the possibility of increased access to nonpublic or currently protected public data sources.

The panel agrees in general with all of the major findings and conclusions in the BTS study:

• The panel concurs with the central conclusion that the present state of available freight data does not permit determination of international trade traffic by state.

• The panel concurs that the untested assumptions relied on by the various models used to calculate the estimates are unlikely to be valid and therefore the resulting estimates are of unknown accuracy.

• The panel concurs that using diesel fuel data to model ton-miles and value-miles of international trade traffic by highway at the state level is unlikely to produce sufficiently reliable estimates for forecasting highway use.

In terms of the underlying purpose of the BTS study, the panel questions the use of value-miles as a useful measure in addition to ton-miles. The panel also questions the separation of estimates of domestic and international trade traffic as being inconsistent with the objective of allocating funds to address needed maintenance for and new construction of the nation's highways. The panel was limited in its evaluation of the utility of the estimates of ton-miles and value-miles for purposes of fund allocation given that such an evaluation necessarily depends on the application with respect to a specific formula, and to date, no formula has been proposed for this purpose.

The panel examined the advantages and disadvantages of various possibilities for additional data collection, including new surveys, additional data collected from administrative records, greater access to existing records, and passive means for data collection using new technologies. Several of these hold promise for both near-term and long-term improvement of the measurement of ton-miles of international trade traffic, but none can now provide reliable estimates of such traffic by state.

1

Introduction

International trade plays a substantial role in the economy of the United States. More than 1.6 billion tons of international merchandise was conveyed using the U.S. transportation system in 2001, representing 10 percent of the nearly 16 billion tons of freight (Bureau of Transportation Statistics, 2004). The need to transport this merchandise raises concerns about the quality of the transportation system and its ability to support this component of freight movement. To that end, Congress directed the Bureau of Transportation Statistics (BTS) to carry out a study under Section 5115 of P.L. 105-178, the Transportation Equity Act for the 21st Century (hereafter referred to as TEA-21). The legislation states (U.S. Department of Transportation, 2003):

> The Director shall carry out a study (1) to measure the ton-miles and value-miles of international trade traffic carried by highway for each state; (2) to evaluate the accuracy and reliability of such measures for use in the formula for highway apportionments . . . ; (3) to evaluate the accuracy and reliability of the use of diesel fuel data as a measure of international trade traffic by state; and (4) to identify needed improvements in long-term data collection programs to provide accurate and reliable measures of international trade traffic for use in the formula for highway apportionments.

> Basis for Evaluation—The study shall evaluate the quality and reliability of measures for use as formula factors based on statistical quality standards

developed by the Bureau in consultation with the Committee on National Statistics of the National Academy of Sciences.

Report—Not later than three years after the date of enactment of this Act, the Director shall submit to the Committee on Environment and Public Works of the Senate and the Committee on Transportation and Infrastructure of the House of Representatives a report on the results of the study carried out under paragraph (1) including recommendations for changes in law necessary to implement the identified needs for improvements in long-term data collection programs.

Under these provisions of Section 5115, the BTS requested the Committee on National Statistics to convene a panel to evaluate the accuracy and reliability of measures of ton-miles and value-miles of international trade traffic carried by highway for each state for use as formula factors for highway apportionments. The specific goals of the study were to review the findings and recommendations of a BTS staff report (Bureau of Transportation Statistics, 2003) prepared in response to Section 5115 and to explore additional data sources and methods for providing improved estimates for use in apportionment formulas.

Since its formation in August 2003, the panel has held two open meetings and two meetings in executive session and commissioned two independent studies. At an initial fact-finding meeting in September 2003, the panel heard presentations from BTS staff and experts who had contributed to the development of BTS's 5115 report (see the agenda, Appendix A). Subsequently, the panel held a workshop in November 2003, at which outside experts and government officials discussed specific aspects of the report (see the agenda, Appendix B). In total, 15 presentations were provided to the panel in open session. The experts provided advice on the validity and quality of data used in developing the report, on alternative and emerging means of developing data on international trade traffic, and on matters of access to public and proprietary data sources (see list of presenters and their topics, Appendix C).

The Panel on BTS International Trade Traffic Study submitted a letter report to BTS in February 2004 that addressed the various conclusions of the BTS report, and several key aspects of the findings (National Research Council, 2004). The letter report assessed the validity of the findings and recommendations in the report prepared in response to Section 5115 of

TEA-21, offered interpretations of its conclusions, and provided suggestions for areas of investigation and analysis.

This final report of the panel represents an expansion of the panel's letter report, going into more depth in several areas, in particular, more fully assessing the validity of the findings and recommendations in the BTS report. This final report draws on workshop presentations and studies commissioned by the committee to address the utility of current data sources, to examine alternative and emerging means for new data collection or increased data access, and to assess the BTS estimation methodology. In addition, this final report of the panel provides suggestions for areas of further investigation and analysis that would improve the analysis contained in the BTS report.

2

BTS Report: Summary of Findings

In response to congressional legislation, the Bureau of Transportation Statistics (BTS) submitted a report to Congress that presented the methods and findings of a study that (1) estimated the ton-miles and value-miles of international trade traffic carried by highway in each state; (2) evaluated the accuracy and reliability of such estimates for use in formulas for highway apportionments; (3) evaluated the accuracy and reliability of the use of diesel fuel as a measure of international trade traffic by state; and (4) identified needed improvements in long-term data collection programs to help provide accurate and reliable measures of international traffic for use in formulas for highway apportionments (Bureau of Transportation Statistics, 2003).

BTS estimated that there were nearly 227 billion ton-miles and $488 trillion value-miles of international trade traffic carried by highways in 1997. Although these are impressive amounts, these estimates, particularly at the state level, are not very reliable. There are no direct measures of ton-miles and value-miles of *international* trade traffic carried by highway in each state. In fact, methods for measuring *total* volumes of ton-miles and value-miles passing over state highways are still in development. Thus, ton-miles and value-miles must be estimated with data from a multitude of public and private sources, none of which was specifically designed for the task.

The quantity of ton-miles is the product of the tonnage of a shipment times the miles the shipment traveled to its destination. The quantity of value-miles is the product of the declared value of a shipment times the miles the shipment traveled. Both ton-miles and value-miles are derived from separate estimates of either weight or value and miles traveled.

Ton-mile and value-mile figures are combined from separate estimates of export and import trade traffic for a variety of categories. Despite growing attention to measures of tonnage, miles, and value, both ton-mile and value-mile estimates have many substantial sources of error. Because the data sources are not designed to directly yield ton-mile and value-mile measures, the estimates require application of model-based estimation techniques that rely on unvalidated assumptions. The models that were used in the BTS study included the Oak Ridge National Laboratory's (ORNL) highway network model to derive estimates of the distances traveled by imports and exports. The accuracy of the distance estimates from this model is unknown. Moreover, data on imports were available only at the state level. The accuracy of the assumptions used to distribute state-level imports to counties within a state is unknown because of a lack of data at the substate level, and substate level allocations are needed to determine the likely route taken, which provides a better assessment of the miles traveled within a state.

Moreover, the various data sources utilized to produce estimates of international ton-miles and value-miles transported by highway have known and unknown errors and biases that affect the accuracy and reliability of the estimates. The errors and biases are due not only to typical sources, including sampling variance, nonresponse variance, and measurement error of various types, but also to various data deficiencies, even though the models mentioned above are used to overcome these deficiencies.

The BTS study identified a number of sources of error for a variety of data sets used to estimate volumes and values for freight arriving here as imports or departing as exports, which we summarize below. In addition, there are shipments involving Canada and Mexico that are not straightforward exports or imports: truck shipments between Canada and Mexico and shipments to or from the United States that are transshipped via Canada or Mexico are not fully captured in current trade data for all modes of transportation, which is also a source of error. Also, some truck transportation was not included in the estimates of the amount of international trade because no reliable data were available, especially including drayage, or short-haul shipments.

STATE-LEVEL EXPORTS

The 1997 Commodity Flow Survey (CFS), resulting from a partner-ship between the U.S. Census Bureau and BTS, is a stratified random sample of commodity flows. The CFS is conducted every 5 years as part of the Economic Census Program, and it is mandatory. The establishments that are surveyed are selected from a frame (a current list) constructed from the Census Bureau's Business Register. For a sample of shipments for a 1-week reporting period each quarter, respondents report the total number of outbound shipments and information on value, weight, primary com-modity, domestic destination or port, airport, or border crossing of exit, foreign destination, and mode of transportation. In terms of shipments, the coverage is extensive: it is estimated that the 2002 CFS gathered information on more than 2.5 million shipments. It is a valuable source of information on the flow of goods by truck and other modes in the United States, and it is the primary source of BTS estimates of ton-miles and value-miles for exports. Unfortunately, the CFS does not provide useful informa-tion for imports since goods are not covered in the CFS until they reach the first domestic shipper covered by the CFS. A substantial amount of travel can be missed before the shipment reaches the first domestic shipper.[1]

While the CFS includes information on shipments of manufacturing, mining, wholesale trade, and selected retail and service industries in the 50 states and the District of Columbia, it excludes shipments of farm-based, forestry, fisheries, transportation, and oil and gas extraction companies. The contribution of these industries to export trade is relatively small, and therefore, while this omission results in the CFS estimates being biased low (excluding other sources of bias), the BTS report says that this bias is a fairly minor source of error. However, CFS also excludes construction, which may represent a more substantial contribution to bias. In terms of all shipments, including imports, only about 60-70 percent of shipments are covered by the CFS.

In addition to sampling variance, the CFS, like any survey, is subject to undercoverage, misresponse, and nonresponse. Not much is known about undercoverage other than that mentioned above. There is no information about the degree of misresponse, which is typically unmeasured. However,

---

[1]The CFS data also support the ORNL's highway network model and are included in public and private databases, such as the Federal Highway Administration's Freight Analysis Framework and the Reebie Associates TRANSSEARCH database.

the sampling variance may have increased of late, since the sample size for the CFS has been reduced from 200,000 in 1992 to 50,000 in 2002. Although this still appears to be a substantial sample size, this reduction could reduce the quality of estimates of international trade traffic since export activity in the CFS is a relatively small percentage of total shipments. Specifically, the BTS report states that exports using all modalities accounted for about 8 percent of estimated total value and 8.2 percent of ton-miles in 1997, and the domestic part of exports shipped by truck accounted for 7 percent of the total value shipped by truck and 5 percent of the ton-miles. The coefficients of variation[2] for the estimates of exports shipped by truck were 5.5 percent for value and 12.4 percent for ton-miles. An analysis of the 1997 response rates indicated an overall unit nonresponse rate of 15 percent, with differences by trade area, size of establishment, and state. Also, the rate of item nonresponse for shipments was 2.7 percent for entries for value and 4.0 percent for entries for weight. Although this amount of unit and item nonresponse likely increases the sampling variance (and therefore the coefficients of variation) of resulting estimates, a more important concern is that this degree of nonresponse could also substantially increase the estimates' bias.

STATE-LEVEL IMPORTS

Imports can be primarily separated into maritime, by land, and by air. With respect to maritime imports, the percentage of imports that arrive at seaports that are subsequently transported by truck is unknown. Also, for those maritime shipments that are subsequently transported by truck, there are only limited data on the state of destination. This lack of direct information necessitates the use of models to estimate the quantities of ton-miles and value-miles for imports transported by highway at the state level.

BTS used the U.S. Foreign Waterborne Transportation statistics to determine the tonnage and value of imports arriving at all U.S. seaports. These data are derived from U.S. foreign trade filings collected by the Customs and Border Protection Agency of the U.S. Department of Commerce. The value of an import is estimated from the weight and the commodity code, which introduces some error. To estimate the amount of maritime imports that are subsequently transported by truck from ports, by com-

---

[2]The standard deviation expressed as a percentage of the estimated quantity.

modity, BTS used data from the Public Use Carload Waybill Sample, collected annually by the Association of American Railroads, to first estimate those quantities for imports transported by rail. (For this purpose, international freights can be identified by an intermodal service code [ISC] and by type of move code.) When rail freight shipments are subtracted from total international freights transported from seaports (ignoring petroleum products), the remainder is assumed to be transported by truck. In some areas, this assumption introduces a substantial bias since it will result in inland and coastwise shipments by water being counted as being transported by truck.

To allocate total imports transported by truck to the state level, BTS used data from the Port Import/Export Reporting Service (PIERS), collected from vessel manifests. Unfortunately, state information is often missing, and the state information may be the address of the importer and not the actual destination. To address this deficiency, BTS sought state-level destination information from the U.S. Foreign Waterborne Transportation System from the Census Bureau, but this request was denied due to confidentiality requirements under Title 13 of the U.S. Code.

For imports arriving by land, BTS used the Transborder Surface Freight Database to collect data on the tonnage, value, origins, and destinations of truck trade with Canada and Mexico. This database contains monthly freight flow data by commodity type and by surface modes for U.S. exports to and imports from Canada and Mexico. Finally, with respect to imports that arrive by air, subsequent transportation by truck is considered negligible, since the distances traveled by truck from airports are typically short.

The BTS report found that, although international trade data are extensive and detailed, the administrative data collected by the Customs and Border Protection Agency and processed by the Census Bureau are limited in their usefulness in measuring international trade traffic on state highways. Critical data elements necessary to identify destinations and transportation modes are not available. Data from the U.S. Foreign Waterborne Transportation Statistics, PIERS, and the Transborder Surface Freight Database provide important pieces of information, but each suffers from issues of data quality and compatibility. Furthermore, the BTS study found that many of the sources of data required to prepare and improve the estimates of international trade traffic by highway by state are collected under pledges of confidentiality and are protected from other use by legal mandate. Also, some of the potential data sources are developed in the private sector and are of unknown quality. As a result of these data problems, BTS used inferior sources of information to produce the requested estimates.

Specifically, the BTS report listed five deficiencies: (1) data on shipment weight for imports transported by truck (which is now required by the Census Bureau as of October 2003); (2) information on the state and county of destination for imports and route taken for exports; (3) information on the port of entry for imports rather than the administrative port; (4) data on all modes of transportation used in shipping to a destination, in addition to the mode used when the cargo arrives in or departs from a U.S. port of entry; and (5) data on truck shipments between Canada and Mexico that transit the United States and shipments to or from the United States that are transshipped via Canada or Mexico. Since BTS does not have access to detailed customs data, it must use extracts published as port-of-entry data in the U.S. Foreign Waterborne Transportation Statistics and trade data with Canada and Mexico in the Transborder Surface Freight Database. Access to customs data would alleviate many, but not remove all, of the five above deficiencies in current data.

COUNTY-LEVEL DATA

Since knowing only the state of destination provides limited information as to the specific highways used, it is important to "carry down" information on import freight movements from the state level to the county level. To do this, BTS initially intended to use four "gravity" models. The four models were of flows from border crossing to counties separately by weight and by value, and flows from seaports to counties, again separately by weight and by value. However, due to computational constraints, the gravity models could not be implemented at the county level and so were only used to re-estimate the ton-miles and value-miles for imports at the state level. Since the estimates already available existed at the state level, the main reason for this application of the gravity model would be to reduce the variance in the available estimates through the replacement of the observations with the fitted values from the model.

To carry these modified estimates of state-level transported freight down to the county level (specifically, the *centroid* of each county), BTS used a model that allocated each state's quantities (ton-miles and value-miles) to counties in proportion to each county's share of total state payroll, as reported in the 1997 County Business Patterns data. (These data include statistics on county-level establishments, employment, and payroll by employment size class for major industry groups and is collected by the Census Bureau.) Given the limited relationship of payroll to international freight, one would expect, at best, that this method would provide only a rough approximation to the true distribution. As a result, there is additional error due to the use of this apportionment method.

ROADS USED

Knowing that freight moved from a port or border crossing to a county centroid is not sufficient for understanding what roads were used in that shipment, and more fundamentally, what mode of transportation was used. For the identification of specific highways used in transporting freight from one county centroid to another, BTS used the ORNL national highway network model, which is part of ORNL's intermodal network model. The model is a geographically based network of major roadways, currently representing 420,000 miles, which simulates shipment routes by roads from two points. The benefit of having this highway model embedded in the ORNL intermodal network model allows one to model, based on transportation costs and travel time, whether freight that is moving from one point to another is likely to be transported using trucks, trains, waterways, or by air. Some of the models that select a mode of transportation or a road use a "least impedance approach" that deterministically selects a transportation option. The remaining models use a logit[3] approach that weights various choices based on their estimated probabilities. The performance of these models is hindered by incomplete data, missing and misaligned roadways, and reliance on a number of questionable assumptions. For instance, the assumption that routes are chosen to minimize costs is violated when truck drivers choose routes based on the need to make multiple pick-ups and deliveries and the ability to mix freight shipments. Also, state specific truck weight limits can affect routes selected. Furthermore, sometimes time of day is factored into a decision in order to avoid congestion. Finally, transportation terminals open and close fairly often, affecting the currency of the ORNL road and modal network. All of these factors contribute to a concern that many of the model's assumptions are invalid. Along the same lines, quantitative measures of the accuracy of the assignments made by the network model do not exist.

---

[3]This is essentially a regression model in which $\log p/(1-p)$ is the dependent variable.

DIESEL MODEL

The third charge to BTS in TEA-21 was to evaluate the accuracy and reliability of the use of diesel fuel data as a surrogate for international trade traffic by state. The focus on this particular approach to modeling international trade traffic is difficult to understand given that no breakdown of diesel fuel data pertains specifically to international trade. Therefore, there is no direct way to use only that portion of diesel fuel data to model international trade traffic by state.

BTS carried out a modification of the required analysis, using diesel fuel data at the state level as the independent variable in a simple linear regression to model the dependent variable ton-miles of highway traffic— not ton-miles of *international* highway traffic. It was hoped that if this model was successful, one could make use of another model that would effectively estimate the portion of diesel fuel used by international trade traffic, possibly by modeling the ratio of international freight ton-miles per state to the total freight ton-miles per state. However, in that case, one would need to rely on an assumption concerning the stability of these ratios, which might not be sensible. Presumably, similar approaches might also be applicable to estimating international value-miles at the state level.

In carrying out the indicated simple linear regression, the percentage of variation explained by diesel fuel sales, R^2, was 88 percent, which represents a correlation between these variables of 0.94. This result was certainly promising, but there were two problems, one of which was pointed out by BTS.

First, while the correlation is quite high, 32 states had differences between the observed and modeled ton-miles that were greater than 15 percent of the observed values, and 11 states had differences of greater than 50 percent. These are important discrepancies that argue against use of these estimates for fund allocation. Yet the discrepancies are not surprising because (1) the fuel sold in a state is not equivalent to the fuel used in a state, (2) there are differences in states' reporting of motor-fuel sales, and (3) the statistics R^2 and the correlation coefficient can be dominated by a minority of larger values in a dataset (e.g., a few large states with high quantities of diesel sales and trade traffic could result in an apparently good fit without providing small relative residuals for many of the smaller states).

Second, this model is intended to be used predictively—that is the regression coefficients would be fixed for several time periods, and R^2 is not a measure of the predictive performance of a regression model. The causal justification for this regression model was very limited. In particular, the regression model failed to account for changes over time in the fuels used to drive trucks and the changes in demand for diesel fuel from other uses.

As requested in TEA-21, BTS was asked not only to provide estimates of ton-miles and value-miles of international highway trade traffic, but also to investigate the reliability of these estimates. In response to this charge, the BTS report asserted that there are important improvements needed in many of the components of their estimation methods for reliably estimating ton-miles and value-miles. Generally speaking, the improvements are needed because of deficiencies, either in current data coverage and quality or with respect to access to data of higher quality that is collected but is currently unavailable.

3

Panel Findings and Conclusions: Data Quality and Reliability

I n this section the panel selectively addresses several major aspects of the Bureau of Transportation Statistics (BTS) report, focusing on the analysis of the requirements for data quality and reliability associated with the purposes of allocation formulas. Section 4 of this report discusses potential (new) sources of data, the strengths and weaknesses of the models used in the BTS report, and the possible role of emerging technologies in obtaining needed data.

The panel concurs with the BTS report that available freight data are not of sufficient coverage and quality to permit precise determination of international trade traffic by state. The panel's review of the availability of validated data to support estimates suggests needed improvements in data collection programs to enhance the accuracy and reliability of international trade traffic measures for formula apportionment purposes.

The panel concurs with the central finding of the BTS study (Bureau of Transportation Statistics, 2003) that the present state of available freight data does not permit precise determination of international trade traffic by state.

The panel was limited in its ability to evaluate the accuracy and reliability of the data for use as formula factors because no allocation formula that depends on ton-miles or value-miles of international trade traffic has been developed or proposed. Without knowing the formula and its purpose, it is not possible to assess how accurate estimates of international trade traffic must be. If, for example, the central purpose of the formula is to compensate states for wear and tear on their highway infrastructure, then even highly precise measures of value-miles will bear little relationship to the objective of the formula, and ton-miles will only roughly reflect the goal of the formula. Less accuracy is needed if the purpose of the formula is insensitive to variations in ton-miles than if the formula is highly sensitive.

More generally, the accuracy that is needed for formula allocation may depend on whether the objective is constructing, maintaining, and operating infrastructure; reducing congestion; minimizing the impact of the highways on the environment; improving the safety of the network; or any of a number of other worthwhile objectives. When evaluating the quality and reliability of data, such factors as error, bias, and transparency would be weighed differently, depending on the objective of the apportionment formula.

It is instructive to compare the quality of international trade traffic data with that of the data used in existing federal highway apportionment formulas. Each major federal highway program has a legislated formula that relies on data obtained from the joint Federal Highway Administration Highway Performance Monitoring System. The primary measures used in the current allocation formulas include lane miles, vehicle miles traveled, diesel fuel data, state population, urbanized area population, and non-highway recreational fuel use. Each of these data elements has its own error measures, as well as an established validation procedure that is documented in guidelines that were carefully developed in a federal-state cooperative venture.

It is important to note that each of the data elements used in the existing formulas is directly and independently measured. The measures are also transparent: their sources are well known, and they use carefully monitored procedures for verification. The current formula allocation factors do not rely on model-based estimation procedures; each has a directly measurable error structure in which bias and variance can be specified.

The characteristics of verifiability and transparency are extremely important for estimates used as allocation factors. Unfortunately, the estimates computed in the BTS report are neither verifiable nor transparent.

The BTS model-based estimates of ton-miles and value-miles of international trade traffic have their genesis in a multiplicity of data sources and methodologies. As the BTS report appropriately notes, "limitations in the data have a negative effect on their accuracy and reliability." Even when it is possible to specify the error associated with one of the data inputs, such as the sampling variance for the Commodity Flow Survey (CFS), it is not possible to specify the quality and reliability of the overall estimates, which reflect the accumulation of the errors[1] of several data sources. It is also not possible to compute the uncertainty that is introduced into the estimates by several of the key methodologies, such as use of the gravity model, the substate allocation procedure, and the highway network assignment procedure.

The BTS report appropriately addresses the need for separately identifying international trade traffic data for use in an allocation formula. However, the panel is concerned that in the long run, a focus on estimating ton-miles and value-miles of international trade traffic by state may distract attention from the larger issues of characterizing sources of highway wear and tear and congestion for purposes of funds allocation. Wear and tear, congestion, and other concerns are not a function of whether the traffic is international or domestic; rather, they relate to such measures as traffic volumes, axle load factors, vehicle types and categories, and weight. It was observed in the panel's workshop that pavement doesn't care whether traffic is international or domestic. Moreover, even total ton-miles are only roughly related to wear and tear and other concerns.

It is likely that total ton-miles of all types of freight traffic moving by state highway can be estimated more accurately than the disaggregated measures of domestic and international trade traffic.[2] Part of the problem of accurately estimating international trade traffic stems from two issues of disaggregation. We offer two examples, on disaggregating diesel fuel use and on computing substate estimates.

---

[1]By errors due to data sources here, we mean all sources by which the estimate fails to equal the true value, which includes various types of bias and sample variance.

[2]The panel notes that the Federal Highway Administration's Office of Freight Management and Operations uses a variety of data sources in the development of the Freight Analysis Framework (FAF), which estimates (as of 1998) and forecasts (to 2010 and 2020) total freight tonnage flows on U.S. highways, as well as U.S.-Canadian truck traffic. The base year of the FAF is the 1993 economic census year, for which CFS data are also available. However, the FAF is assembled from existing data sources. Therefore, it is subject to the same problems of low frequency of acquisition and questionable accuracy as the data sources.

Disaggregating Diesel Fuel Use. The BTS report assesses the accuracy and reliability of diesel fuel data as a measure of ton-miles of international trade traffic on state highways. Diesel fuel data is already a factor in the apportionment of funds for the National Highway System program, but the measure has several weaknesses when pressed to serve as data for allocation purposes. It is difficult to estimate the number of gallons consumed for road use in each state because fuel may be purchased in one state and used in another, and there are differences in state reporting of motor fuel sales. Procedures used to adjust sales data to use data (by means of reports submitted by carriers under the International Fuel Tax Agreement) add variability to the estimates. Furthermore, not all diesel fuel is used for on-the-road vehicles: for example, diesel fuel is also used to run construction equipment and railroad locomotives, and therefore any heterogeneity in these uses across states would further limit the utility of diesel sales for this purpose. Other factors further reduce the reliability of these estimates. Nonetheless, the diesel fuel data are currently used as a factor in the apportionment formula. The BTS report states that these estimates "inherently capture international trade traffic."

Several computations are necessary to separate international from domestic diesel fuel use. The BTS report uses a regression analysis of the relationship between the estimate of *total* ton-miles by truck by state and diesel fuel sales by state to produce a formula for predicting ton-miles based on diesel fuel use. The report finds that the regression relationships for many states are weak, suggesting that "diesel fuel usage estimates do not provide good predictors of *total* trade ton-miles on a state-by-state basis." The panel is concerned that the next step—to break down the weak total relationships into domestic and international components—would require comparing estimates of international trade traffic ton-miles from the BTS study with estimates of total ton-miles by state from other sources.

The panel concurs with the BTS report that the process of disaggregating diesel fuel use into its domestic and international trade traffic components would compound variability in that it would apply a ratio of two estimates with significant variability with state diesel fuel use data that have known quality problems.

Computing Substate Estimates of Import Trade Traffic Flows. Existing data sources provide estimates of imports flowing to each state, but not of the distribution of imports within states. The distribution of imports within a state is estimated by assuming that imports to each county are proportional to annual payrolls computed from data in the 1997 County Business Patterns. The accuracy of this assumption is unknown. For discussion of this point, see National Research Council (2004).

The issue of the quality of data used to support fund allocation formula programs is a complicated one; for a broad review of the considerations involved and much more general issues concerning inputs to and use of fund allocation programs, see National Research Council (2003b).

4

Data Sources, Models Used, and Emerging Technologies

POTENTIAL DATA SOURCES

B ecause there are no direct measures of ton-miles and value-miles of international trade traffic carried by highways for each state, data from a variety of sources, none of which was specifically developed for this purpose, have to be used to provide estimates. Previous National Research Council studies have assessed several of these data sources (see, e.g., National Research Council, 2003a).

The congressional legislation that mandated the Bureau of Transportation Statistics (BTS) study implicitly recognized many of the difficulties that would be encountered in developing measures of ton-miles and value-miles of international trade traffic carried on highways in each state, and it directed BTS to identify long-term data improvements to provide accurate and reliable measures for use in highway apportionments. This section provides a few observations about the potential of various approaches to sufficiently improve data to enable its use in formula allocation.

A prior BTS report (Bureau of Transportation Statistics, 2003) listed four general deficiencies in the current data sources: (1) there are international shipments involving Canada or Mexico that are not exports or imports;[1] (2) short-haul shipments are not included; (3) information on all

---

[1] We note that this deficiency may not be due to data limitations—it may be definitional.

modes of shipment using multiple modes is often missing; and (4) complications can result from situations in which the country of origin can change when the goods are substantially transformed prior to end use.

For data on exports, the BTS report also identified five deficiencies in the Commodity Flow Survey (CFS): (1) it excludes a number of industries; (2) it has substantial nonresponse and sampling variance for exports; (3) it is only collected every 5 years; (4) for over 30 percent of its export records, the port of exit is missing; and (5) it does not provide useful information on imports.

For data on imports, the BTS study listed five deficiencies: (1) the value of imports must be estimated from commodity code and weight; (2) truck traffic from maritime imports must be estimated by the total minus imports transported by rail, which ignores petroleum products and inland and coastal shipments by water; (3) allocation to the state level uses Port Import/Export Reporting Service (PIERS) data, for which the state of destination is often missing; (4) the state may be the address of the importer and not that of the destination; and (5) truck traffic from imports arriving by air is ignored.

Finally, the BTS report pointed out two problems in estimating transportation at the substate level: there is datedness and a lack of focus on foreign business in the County Business Patterns data, and there are missing values of various kinds in calculating the Oak Ridge National Laboratory's (ORNL) highway network and the intermodal network.

To address these deficiencies, the BTS report listed five data needs: (1) data on shipment weight for imports transported by truck, to eliminate the need to estimate this by subtraction; (2) information on state and county of destination for imports and information on county of destination for exports, to eliminate the need to use PIERS data and the county allocation model; (3) information on the port of entry rather than the administrative port; (4) data on all modes of transportation used in shipping to a destination, in addition to the mode used when the cargo arrives in or departs from a U.S. port of entry; and (5) data on in-transit and transshipments.

The BTS report mentioned three possibilities for (additional) data collection that we review in this section: (1) obtaining access to currently restricted data; (2) expanding the information requested on administrative records; and (3) initiating a new survey. In addition, there are a number of emerging technologies that may be useful for this purpose; we consider them in the last part of this section.

Obtaining Access to Currently Restricted Data

The Census Bureau interprets Title 13 provisions on data privacy to extend to data from the Customs and Border Protection Service. We are not fully aware of the underlying reasons for this ruling but it may be that some compromise can be arrived at through one of the following avenues: (1) requesting use of an extract at some aggregate level or on a sample basis with all identification of the ownership omitted to help ensure privacy; (2) using various analytic techniques for disclosure protection, such as noise incorporation; (3) providing the software for analysis to the Census Bureau and requesting that the Bureau provide the aggregate estimates at the county level in return, observing disclosure protection; or (4) having BTS personnel work in the Census Bureau as special sworn employees in a restricted-access environment.

Expanding the Information Requested on Administrative Records

The possibility that PIERS data could include collection of data on shipment value and the state of destination rather than the state of the importer should be explored. In addition, the cause of missing data on state of destination should be examined and actions taken to reduce the occurrence of missing data.

International trade statistics are collected by the Customs and Border Protection Agency for most U.S. international transactions and are processed by the Census Bureau's Foreign Trade Division. This database is large, containing more than 3 million import and 2 million export transactions per month (Monk, 2003). If the records in this database included a few additional key data elements, they could be used directly to report international trade volumes, values, destinations, and modes. However, the data are largely administrative in nature, provided by shippers, not carriers. The records contain value data but no weight data; they have limited information on geographic location of origin and destination; they have uncertain assignment of port geography; and they fail to capture all modes of transportation. The export statistics are of particularly questionable quality since they receive little agency scrutiny. Estimates suggest that undercoverage of exports is between 3 and 10 percent. All of these factors tend to limit the utility of these data for apportionment formula purposes.

The panel observes that there are several fortuitous projects under way, which, in the long term, hold promise of filling these information gaps.

The Custom and Border Protection Agency is streamlining and enhancing the international transaction data systems under the Automated Commercial Environment (ACE) system, a $1 billion program. This upgrade is linked to a government-wide program that would meet the needs of more than 100 agencies for trade data. Over the next several years, more than $100 million will be invested in the International Trade Data System (ITDS) to modernize the data and access to them (Fiocco, 2003). This combined effort holds promise, in the long term, of filling information gaps relevant to the estimation of international highway trade traffic.

The ITDS could provide a "one-stop" filing system for both carriers and shippers through use of an Internet-based integrated government-wide system of collection of trade and transportation data. Such a system could meet the needs of all federal agencies with responsibilities related to international trade. Under ACE and ITDS data will be reported from both carriers and shippers, with a unique identifier associated with each international transaction.

The plan is for ITDS/ACE to become the government's front-end information technology system for all federal trade and border agencies, providing more accurate, thorough, and timely data on imports, exports, and in-transit information on cargoes, conveyances, and crews. The hope is to use these data to improve compliance with and enforcement of trade requirements, reduce the cost and burden of processing trade transactions, facilitate interagency coordination, develop and distribute higher quality and more comprehensive and timely information on U.S. trade, and improve Department of Transportation access to this information. Integration is extremely desirable since there are more than 100 agencies and offices involved in trade, and as a result, some data collection has been redundant, and data collection currently makes use of some paper forms and incompatible automated systems. ITDS and ACE will also promote standardization of federal international trade and transportation requirements.

There is certainly a tension between information and security, so it is not surprising that while ITDS and ACE may have the promise of resolving most of the data gaps discussed in this report, access may be restricted. The panel hopes that the proposed strategies for dealing with Title 13 issues concerning current Customs and Border Protection Agency data can be addressed in time to make use of this exciting new data collection and coordination effort.

Expanding the CFS or Initiating a New Survey

New survey data collections are expensive, and must therefore be very carefully considered, especially given that there is a relevant survey, the Commodity Flow Survey (CFS), that is already in place. Since the CFS samples domestic exporters, one might consider augmenting the CFS with a sample of businesses that import into the United States. However, there would be substantial complications in developing a relevant frame, and the mandatory status of the CFS cannot be extended to foreign importers. A pilot study allowing analysis of the quality of the information collected might be informative and relatively inexpensive. In addition to adding information on imports, the CFS might also be examined to see whether the addition of more contextual information to address some of the data deficiencies discussed above could be accommodated without affecting the quality of the data already collected.

Barring changes to the CFS, instituting a new survey raises the possibility of including data elements that are not required by the Customs and Border Protection Agency (and therefore will not be available in the future on ITDS/ACE). Two options for such a survey are a shipper-based survey or a carrier-based survey. These surveys would need to provide timely data, which may mean an annual survey, though that remains to be determined through analysis of pilot data.

As noted by BTS (U.S. Department of Transportation, 2003), a carrier-based survey might involve surveying a sample of trucks entering or leaving major seaports and airports and therefore possibly engaged in international trade. (Unfortunately, carriers may not know the final destination of the cargo they carry.) The resulting data would need to be reconciled with port-specific reporting of international freight cargo. As noted in that report, in the mid-1970s the Census Bureau had access to all foreign trade paper documents collected by the Customs and Border Protection Agency, and those documents served as a universe from which a follow-up survey of shipper-based establishments was conducted. Today, an analogous survey would involve sampling international transactions filed with that agency through ACE. These surveys collected reliable data on inland destinations of imports and origin of exports, the mode of transportation used, and the weight of the cargo. Those follow-up surveys have not been conducted since the mid-1970s.

MODELS USED

Three models were used in the production of the BTS estimates of ton-miles and value-miles of international highway freight: the gravity model, a state-to-county allocation model, and a road selection model. We provide short descriptions of these models and include some discussion of problematic aspects of each modeling application.

Gravity Model

The model initially selected by BTS to "carry down" state-level estimates of value-miles and ton-miles to the county level is known as the gravity model. A gravity model is a method for estimating the flow of goods or persons from one location to another. There are many different versions of the gravity model. The one used for the BTS analysis has the form $T_{ij} = A_i B_j F_{ij}$, where T_{ij} is the flow (measured, for example, in tons or value per year) from state i to state j, A_i is a state-specific parameter that represents the total tons or value originating in state i regardless of destination, B_j is a state-specific parameter that represents the tons or value arriving at state j regardless of origin, and F_{ij} represents the "impedance" to flow between i and j. F_{ij} is typically a composite of the cost and travel time required to ship a commodity from i to j. In the model used for the BTS study,

$$F_{ij} = \exp\left(\sum_k \theta_k c_{ij}^k\right)$$

where c_{ij}^k (k = 1,2,..., K) represents impedance-related variables, such as travel time and cost, and the θ_k's are constant parameters. The A_i's, B_j's, and θ_k's are estimated by fitting the model to data on flows between states (see Metaxatos, 2002 for further details). Because in the BTS application the model was estimated at the state level, it cannot be used directly to estimate flows at a finer level of geographical aggregation (e.g., county-to-county flows).

Although the gravity model could have been applied at the level of county centroids, it would have required that the factors related to the costs of transportation from one port or border crossing or county to another be used, which was computationally too demanding (involving a matrix of dimension of more than 3,000 by 3,000). Therefore, the gravity model was applied at the level of states, which is the level at which the estimates already existed.

The panel is unconvinced that the gravity model plays a useful role in the estimation of ton-miles and value-miles for international trade traffic by highways. Replacing the observed measures with their estimated expected values (given the gravity model) in some applications could reduce variances without appreciably increasing bias. But to do so, the model form would have to be justified either by subject-matter considerations or empirically, or preferably both, since otherwise model misspecification would likely eliminate any variance-reduction benefits from the use of estimated expected values in replacing the observed measures.

A second possible justification for the use of the gravity model is that one might be able to use the model for forecasting subsequent years' highway use. However, there has been no suggestion for how the origination and destination factors could be updated for future time periods. This lack is further complicated by the introduction of new ports or border crossings over time, as well as various kinds of state or county dynamics resulting from changes in economic circumstances, user charges, various technological advances, etc. None of this is explained by the gravity model.

State-to-County Share Model

Using states as the geographic level of analysis does not provide very accurate assessments of distances of trade traffic, since knowing the destination state often does not specify the highways used. Therefore, it is extremely important to distribute the shipments to the level of counties (county centroid) to obtain more information about roads used.

BTS used the Census Bureau's County Business Patterns data to distribute the freight movements from the state to the county level. To do this, BTS used the ratio of each county's total payroll to total state payroll to allocate that proportion of the state total shipments to the individual counties. (This process is adjusted to ignore establishments that do not have a county identification.) Clearly, this model is likely to be subject to substantial misspecification given its use of a single variable that is only weakly related to the transportation of freight, let alone international freight.

County Centroid Route Assignment Model

Given the identification of the destination county centroids, one has to identify the routes taken by trucks from seaports and border crossings to these centroids: this is accomplished through use of the ORNL highway network assignment model, which determines the distances traveled essentially through selection of the least cost/time route. In addition, one also has to identify the mode of transportation when there are alternatives to traveling by road. One important benefit of the ORNL highway network is that it is embedded within the ORNL intermodal network, which permits one to model the mode and route used to move freight from one node in the network to another on the basis of travel costs and time required. This assignment model, within the same general approach, sometimes makes use of a least impedance approach that deterministically selects a transportation option and sometimes uses a logit approach that weights various choices based on their estimated probabilities of selection. As far as the panel is aware, neither the least impedance nor the logit approach has been supported by empirical validation. Unfortunately, this overall approach is hampered by incomplete data in several respects. In addition, there is no guarantee of the uniform applicability of the assumption that the movement of freight always conforms to an economically optimal selection of routes.

The ORNL highway model produces variances, accompanying each origin-destination pair, that are estimated using the bootstrap method.[2] Neither this assignment model nor its associated variances, to the panel's knowledge, have been empirically validated with actual data on route selection.

Model Validation

In all three applications of models to produce the BTS estimates, there is a lack of model validation, both external and internal. External validation is the comparison of model forecasts with realized values. Internal validation is the understanding of the performance of a model through examination of the functioning of individual components, especially how variability propagates through a model. Model validation is essential both to understand the current performance of a model and to direct efforts efficiently for model improvement in the future.

Model validation has benefited from enormous progress in the last 25 years, but this research is not reflected in the BTS work. (Some of the

---

[2]The bootstrap is a technique for estimating variances in complicated situations that approximates sampling from the unknown population by sampling instead from the observed data (see, e.g., Efron, 1982).

newer methods for model validation can be found in Morgan and Henrion [1990], National Research Council [1991], and Mulry and Spencer [1991].) In an effort to validate the gravity model, BTS computed a chi-square goodness-of-fit statistic to assess how well the gravity model fits at the state level.[3] However, this statistic provides little information about the performance of the model in the context of prediction. More importantly, it does not indicate whether use of the gravity model gives smaller errors in the estimated origin-to-destination flows than does use of the flow data that were used as inputs to estimation of the model.

For the assignment model, ORNL has carried out a number of careful assessments of the quality of the data used to construct the networks that are relied on, but there has been no validation that the assignments produced agree with those selected in practice by freight movers, as mentioned above, and no sensitivity analyses of the effects of missing data on the assignments has been provided.

In concluding this section, the panel notes one related point. Another approach to providing estimates of higher quality for states, other than additional data collection, would be to try to use combining-information models with variables that are collected more frequently than the current survey frequencies, possibly at lower levels of geographic aggregation. Making full use of any time-series structure in these independent variables might support projections several years in the future, in addition to making current small-area estimates. We have not carried any research out in this area, but we believe there is some promise for substantially improving the current estimates through use of these methods.

EMERGING TECHNOLOGIES

One problem with many transportation surveys is that they require time by truckers to respond, which results in substantial noncompliance

---

[3]In the BTS study, T_{ij} is interpreted as the expected value of the actual flow from i to j, N_{ij}, which is treated as a random variable. Thus, $T_{ij} = E(N_{ij})$. T_{ij} is unknown and is estimated from the fitted gravity model. Denote the estimate by $\hat{T}_{ij}$. Then the chi-square test statistic is

$$\sum_{i,j} \frac{\left(N_{ij} - \hat{T}_{ij}\right)^2}{\hat{T}_{ij}}$$, with degrees of freedom equaling the total number of terms in the sum minus the number of origin factors, minus the number of destination factors, minus the number of cost factors −1.

and incomplete reporting. There are a variety of emerging technologies and techniques that might be used to more unobtrusively capture information on weights, value, and vehicle types, including weigh-in-motion and radio frequency tag technologies that would identify vehicles and cargos while in motion. There are also important advances in traffic counting technology. As described in this section, these emerging technologies that are now either in advanced development or in the early stage of deployment may soon enable more informed estimates of the amount and value of international trade traffic on state highways. The "answer" will most likely consist of a combination of technologies that provide information on both weight and cargo. It is possible to envision a system of weigh-in-motion and tandem radio frequency technologies (see below) that tie directly to emerging company and customs databases. Also, the Canadian roadside sample program[4] may inform the selection of the vehicles to sample for data collection.

However, none of the current technologies can currently provide reliable estimates of ton-miles and value-miles of international trade traffic. Furthermore, characteristics of trade, especially in origins and destinations, are unlikely to be addressed by technological advances in passive information collection.

In order to take advantage of many of these emerging technologies, an up-to-date, accurate geographic information system (GIS) coverage will be needed. Many states and local governments are using their own local GIS to locate and perform maintenance on their intelligent transportation systems equipment. Electronic maps with the geocoded location of these infrastructures would need to be "fused" across county and state boundaries to build a national network. The underlying infrastructures (including the devices producing the data streams) are dynamic and are intended to grow over time, which would require a dedicated, national effort to ensure accuracy and timeliness of the locations of these data generators. Mobile data sources (e.g., global positioning system [GPS] units on trucks) also need to be identified spatially. Without advances in the ability to upgrade and maintain the necessary GIS data, many of the potential data sources cited in this report will remain underused.

The rest of this section briefly discusses eight emerging technologies:

---

[4]The Canadian roadside sample program is an intercept survey of commercial trucks at U.S.-Canada border crossings.

- weigh-in-motion;
- virtual weigh station;
- electronic screening, including radio frequency identification and dedicated short-range communications;
  - smart trucks, which use global positioning and similar systems;
  - bar codes;
  - video image detection;
  - vehicle and card inspection; and
  - archived data user service.

For each technology, each description is followed by comments on its advantages and disadvantages, cost, and likelihood of deployment.

Weigh-In-Motion Technology

Weigh-in-motion (WIM) technology is the process of measuring the weights of axles on trucks while in motion. Various types of WIM technology are capable of highly accurate measurements for a particular speed. The more expensive WIM systems can accurately measure weights at "highway" speeds (± 10 percent accuracy for 95 percent of the trucks), while less expensive systems are accurate at slower speeds (± 15 percent accuracy for 95 percent of the trucks).

Advantages and Disadvantages. The advantage of WIM technology is that it will allow the capture of fairly accurate truck weights, and it requires no personnel to operate. The disadvantage of WIM is that not every road is outfitted with WIM technology. Also, WIM can only provide weights—it cannot provide any commodity information that can be used for estimating value.

Cost. The costs for WIM technology depend on the type of technology used. The initial equipment cost for piezoelectric technology can be as little as $2,500/lane, and the cost of single load cell can be as high as $39,000/lane.[5] While initial costs vary greatly, the overall life-cycle costs are much more equivalent.[6]

---

[5]International Road Dynamics, Inc. (2001).

[6]Recent experience by CalTrans suggests that the current cost estimates of WIM might be too low, since full-speed WIM sensors require a huge investment in high-quality pavement to ensure that the truck does not bounce over the sensors. Similar unexpected cost considerations might apply to other estimates presented here, given the uncertainties that accompany new technology.

Likelihood for Deployment. Near Term: many jurisdictions are deploying WIM technology today and are seeking funding to add more WIM sites with plans for enforcement purposes.

Virtual Weigh Station (VWS) Technology

Virtual weigh station technology uses WIM, closed circuit television cameras, and, in some cases, optical character recognition technology (for license plate numbers). VWS allows jurisdictions to identify trucks, collect their weight data, identification data, and also their speed. The system is also used to capture data on overweight trucks.

Advantages and Disadvantages. The advantage of VWS technology is that it can not only capture weight data, it can also capture speeds, and identify the trucking company and the specific vehicle (by license plate). Another advantage of VWS technology over a WIM-only site is that the image of the truck can indicate to the jurisdiction the commodity given sighting of, say, a logging truck, a car hauler, a tanker truck, or some other obvious kind of carrier. The disadvantage of this technology is that there is no information on the contents of a closed trailer, that is, the value—only the weight.

Cost. The cost for VWS technology is approximately $150,000-$200,000 per site. This includes WIM technology, cameras, and computer storage equipment (and possibly a method for communicating the data to an enforcement vehicle or a weigh station site).

Likelihood for Deployment. Near to mid-term: several jurisdictions are deploying VWS sites today; many others have VWS in their jurisdiction's technology deployment plans.

Electronic Screening

Radio frequency identification (RFID) tags or dedicated short-range communication (DSRC) systems use radio frequency to communicate between a truck-mounted tag and a roadside tag reader. This communication link has many applications, one of which is to identify freight as it is loaded and offloaded from trucks. The unique ID number of the tag corresponds to information contained in a database that houses detailed information on the freight with which the tag is associated. This two-way communication link is used primarily on the interstate highway system to identify a specific truck (or company), run a quick database check to determine if the truck is in compliance (paid taxes, etc.), and it can also alert a driver to pull into a weigh station or port-of-entry to be inspected.

Advantages and Disadvantages. The advantage of using RFID technology is that it can identify the motor carrier as the truck travels down the highway. If a jurisdiction has a database link to the truck and its manifest (most do not have this capability), a great deal of information can be obtained. The disadvantage of DSRC in this application is that it can only directly identify the carrier and the truck. There is no commodity information, nor is there any weight information for many DSRC sites. Another disadvantage is that only a very small percentage of trucks carry DSRC tags and participate in the electronic screening programs. Moreover, the various programs that do exist are not interoperable so a tag that can be identified in one state may not be recognized in another state, and the data are not a public record that can be used to track commodity flows by public agencies.

Cost. The cost for DSRC screening technology is approximately $25-$75 (personal communication) per tag. Depending on the program the carrier enrolls with which the carrier may or may not have to purchase the tag.

Likelihood for Deployment. Near term and ongoing: several jurisdictions are deploying electronic screening programs across the country, but participation is low.

Smart Trucks

Smart trucks refers to trucks that are equipped with such technologies as GPS, satellite communications, and on-board diagnostic and monitoring capabilities. These trucks have the capability of monitoring, reporting, and performing automated functions while the truck is in motion. For example, sensors may indicate that the trailer doors have been opened and send a message by satellite, or sensors may weigh the contents of the trailer and transmit this information to dispatch for billing purposes. These systems also have the capability to monitor engine performance: for example, they can allow monitors in central locations to conduct such operations as adjusting the air-fuel ratio while the truck is in motion without the driver's knowledge; they can even disable the vehicle in case of theft. These systems can also locate a vehicle and determine if a truck has detoured from its scheduled route.

Advantages and Disadvantages. The advantage of having smart trucks is constant communication with a motor carrier's assets. The cargo can be monitored for temperature, the engine's performance can be observed, and vehicle speeds and braking applications can be monitored and reported. Also, the driver's logs can be maintained electronically. The disadvantage, in the context of this report, is that this information is private: unless the motor carrier volunteers this information, a government jurisdiction will have difficulty obtaining it. And as with other systems, none of the information, operating conditions, location and freight security and temperature monitoring provides value information for the cargo.

Cost. The costs for GPS tracking systems can start as low as $1,100 for commercial fleet tracking applications, but this does not include software.[7] There are both less expensive and much more expensive options for tracking vehicles. Many of the features discussed above involve the integration of various technologies linked to satellite communications systems. The cost for this technology varies depending upon the complexity of the system. A system with pallet readers at two locations will be considerably cheaper than systems with loading and unloading facilities in every state across the nation. A typical tag for this application can cost as little as $1.00 per tag.[8]

Likelihood for Deployment. Near term and ongoing: several of the larger trucking companies are deploying this technology today; others are waiting for the benefits to outweigh the costs. Legislation is being developed that will allow tax incentives for the purchase of such technologies that can increase safety and security.

Bar Codes

Bar code readers are a common technology in freight tracking and many other industries. This communication link has many applications, one of which is to identify freight as it is loaded and offloaded from trucks. The unique bar code assigned to a package corresponds to information contained in a database that houses detailed information on the freight with which the tag is associated.

---

[7]See Terratracker (2004) available: http://terratracker.com/html/fleet.html [Accessed August 27, 2004].

[8]Doug Argie at Peak Direct, authorized distributor for Zebra Technologies, personal communication.

Advantages and Disadvantages. The advantage of using bar code technology is that it can provide detailed information for each parcel loaded on a truck and that data can follow the freight from pick-up to delivery. The disadvantage of this technology is that the information is not a public record that can be used to track commodity flows by public agencies.

Cost. The cost for this technology is relatively inexpensive but depends on the complexity of the system: Hand-held units can range between $200 and $3000[9] depending upon the technology. However, with the widespread use of bar codes, the cost of this technology is substantially lower than a closed proprietary system.

Likelihood for Deployment. Short term: This technology is widely used today by package delivery companies, grocery stores, etc.

Video Image Detection

Video image detection systems use machine vision technology to compile and analyze traffic data collected with closed circuit television (CCTV) systems. Video image detection can be used to monitor freeway conditions, capture speeds, count vehicles, and classify vehicles.

Advantages and Disadvantages. One advantage of video image detection is that no additional equipment is required from the freight community. Also, the true counts of commercial vehicles can be obtained. In addition, limited commodity information can be collected such as logging trucks, car carriers, tanker trucks, etc. A disadvantage is that weights cannot be collected, and there is also no way of identifying the contents of closed trailers.

Cost. The cost for this technology can be relatively high, with color CCTV cameras costing from $10,000 to $50,000 and annual maintenance costs ranging from $200 to $1,000. In addition, one needs a communications link, software systems, and algorithms for automated surveillance (Loukakos, online).[10]

Likelihood for Deployment. Mid- to long-term: there are more than 5,000 video image detection systems now in operation, but most of these systems are outside of the United States.

---

[9]System ID Warehouse: Available: http://www.systemid.com/BARCODE_SCANNERS/ ?source=overture [Accessed August 27, 2004].

[10]"Video-Image Detection" Dimitri Loukakos, UC Berkeley and Caltrans; Available: http://www.calccit.org/itsdecision/ [Accessed August 27, 2004].

Vehicle and Cargo Inspection System

A vehicle and cargo inspection system (VACIS) technology uses gamma radiation to essentially take a "picture" of the contents of vehicles. This technology has located and identified drugs, people, and also identified cargo that was not listed in a cargo manifest.

Advantages and Disadvantages. The advantage of VACIS technology is its ability to see the inside of vehicles without removing their contents. It has proven effective for security purposes in many locations for trucks as well as passenger vehicles. The disadvantage of this system is that inspection speeds are very slow, requiring trucks to stop; in practical terms, this means that not all trucks can be inspected.

Cost. The cost for the VACIS machine is approximately $1M per unit.[11] This technology is available for permanent installation locations as well as temporary setups used to inspect individual pallets of freight.

Likelihood for Deployment. Mid-term: there are limited deployments of the VACIS technology today for roadside, ports, and railroads. However, in today's security environment, more deployments in the near term are likely.

Archived Data User Service

Archived data user service (ADUS) is the storage of transportation-related data, for a given period of time, for use by anyone for planning purposes, trend identification, etc. These data are primarily collected through means of intelligent transportation systems. Many of the technologies mentioned in this report can support an archived data user service.

Advantages and Disadvantages. An advantage of using ADUS is that there is a great deal of information collected today that is purged at set intervals. Through ADUS, these data could be collected and used for freight movement data, and possibly also for origin-destination data. The major disadvantage is that most of these data are lacking in specific information for a given vehicle. For example, video, WIM, or other technologies can only identify a truck—not which truck.

Cost. The cost for ADUS is hard to quantify. Some systems already have ADUS capabilities built in, so the cost is nil. Other systems may

---

[11]Brian McDaniel, Science Applications International Corporation, San Diego, personal communication.

require modifications to manage such large amounts of data—including scrubbing the data and supplying notes on data gaps or documenting when data collection systems are out of calibration or offline due to maintenance, etc.

Likelihood for Deployment. Mid-term: although some ADUS systems are currently up and running, ADUS standards are still in development. Lately, more information technology procurements are including ADUS requirements.

References

Bureau of Transportation Statistics

2003 *Measurement of Ton-Miles and Value-Miles of International Trade Traffic Carried by Highway for Each State*. Washington, DC: U.S. Department of Transportation.

2004 *International Trade Traffic Study: Transportation Equity Act for the 21st Century*. P.L. 105-178, section 5115. Preliminary Report. Washington, DC: U.S. Department of Transportation.

Efron, Bradley

1982 *The Jackknife, the Bootstrap, and other Resampling Plans*. Philadephia: Society for Industrial and Applied Mathematics.

Federal Committee on Statistical Methodology

1978 Statistical Policy Working Paper 3-An Error Profile: Employment as Measured by the Current Population Survey, Office of Federal Statistical Policy and Standards. Prepared by Camilla A. Brooks and Barbara A. Bailar, Subcommittee on Nonsampling Errors, Federal Committee on Statistical Methodology, U.S. Department of Commerce.

Fiocco, M.J.

2003 Overview of International Trade Date System (ITDS) & Automated Commercial Environment (ACE). Paper Presentation to the International Trade Traffic Study Workshop, November 21, 2003. Unpublished paper, U.S. Department of Transportation.

International Road Dynamics, Inc.

2001 *Weigh-In-Motion Technology Comparisons: Technical Brief.* Saskatoon, Saskatchawan: Author.

Loukakos, Dimitri

2004 Video-Image Detection. Available: http://www.calccit.org/itsdecision/ [Accessed August 27, 2004].

Metaxatos, Paul

2002   *International Trade Traffic Study (Section 5115): Measuring the Ton-Miles and Value-Miles of International Trade Traffic Carried by Highway for Each State: Estimation of Origin-Destination Flows and Confidence Intervals: Final Report.* Washington, DC: U.S. Department of Transportation.

Monk, C. Harvey, Jr.

2003   Access to Export/Import Trade Data. Presentation to the International Trade Traffic Study Workshop, November 21, 2003. Unpublished paper, U.S. Census Bureau.

Morgan, M. Granger, and Max Henrion

1990   *Uncertainty: A Guide to Dealing with Uncertainty in Quantitative Risk and Policy Analysis.* Cambridge, England: Cambridge University Press.

Mulry, Mary, and Bruce D. Spencer

1991   Total error in PES estimates of population: The dress rehearsal census of 1988. *Journal of the American Statistical Association* 86:839-854 (with discussion 855-863).

National Research Council

1991   *Improving Information for Social Policy Decisions: The Uses of Microsimulation Modeling: Review and Recommendations.* C.F. Citro and E.A. Hanushek, eds. Panel to Evaluate Microsimulation Models for Social Welfare Programs, Committee on National Statistics. Washington, DC: National Academy Press.

2003a   Letter Report on the Commodity Flow Survey. Committee to Review the Bureau of Transportation Statistics (BTS) Survey Programs. Transportation Research Board, Washington, DC: The National Academies Press.

2003b   *Statistical Issues in Allocating Funds by Formula.* Panel on Formula Allocations, Thomas Louis, Thomas B. Jabine, and Marisa Gerstein, Eds., Committee on National Statistics, Division of Behavioral and Social Sciences and Education. Washington, DC: The National Academies Press.

2004   Letter Report to the Bureau of Transportation Statistics. Committee on the Bureau of Transportation Statistics International Trade Traffic Study. Committee on National Statistics, Division of Behavioral and Social Sciences and Education. Washington, DC: The National Academies Press.

System ID Warehouse, Inc.

2004   Barcode Scanners. Available: http://www.systemid.com/BARCODE_SCANNERS/?source=overture [Accessed August 27, 2004].

Terratracker

2004   How Secure Is Your Fleet? Available: http://terratracker.com/html/fleet.html [Accessed August 27, 2004].

U.S. Department of Transportation

2003   Measurement of Ton-Miles and Value-Miles of International Trade Traffic Carried by Highway for Each State. Washington, DC: U.S. Department of Transportation.

Appendix
A

Agenda for Open Sessions

INTERNATIONAL TRADE TRAFFIC STUDY

FIRST PANEL MEETING,
WOODS HOLE, MASSACHUSETTS

THURSDAY, SEPTEMBER 18, 2003

9:30-10:00AM   Scope and Tasks for the Panel
                Joel Horowitz

10:00-10:30AM   Origins of the 5115 Report
                Bill Bannister, Bureau of Transportation Statistics

10:30AM-12:00PM Presentation of Findings of the BTS 5115 Report
                Ho-Ling Hwang, Oak Ridge National Laboratory
                Paul Metaxatos, Universal Technical Systems, Illinois

12:00-1:00PM   Lunch

1:00-2:00PM   Discussion of Data Sources on Ton-Miles and Value-
                Miles of International Trade Traffic
                Amelia Regan

2:00-2:45PM   Discussion of Statistical Models with Emphasis on the
                     Use of the Gravity Model
                     Hani Mahmassani

2:45-3:15PM   Break

3:15-4:15PM   Discussion of Survey Data Quality Issues (with
                     Emphasis on the Commodity Flow Survey)
                     James Lepkowski

4:15-5:00PM   Discussion of State Data Sources
                     Ron Tweedie

5:00-5:45PM   Discussion of Allocation Formula Issues
                     Bruce Spencer

5:45-6:00PM   Summary of Issues
                     Joel Horowitz

Appendix
B

Workshop Agenda

INTERNATIONAL TRADE TRAFFIC STUDY

NOVEMBER 20-21, 2003
ROOM 110
GREEN BUILDING, 2001 WISCONSIN AVENUE, N.W.
WASHINGTON, DC 20048

THURSDAY, NOVEMBER 20, 2003

9:30AM-12:00PM   Session I: ***Concepts and Definitions***
          Amelia Regan, Moderator
          Conceptual Underpinnings of Ton-Miles and Value-Miles
          Irwin Silberman, Bureau of Transportation Statistics
          Missing Pieces in Ton-Mile Estimation and ORNL Highway Network
          Frank Southworth, Oak Ridge National Laboratory

12:00-1:00PM   Lunch

1:00-2:15PM   Session II: ***Allocation Formula***
          Bruce Spencer, Moderator
          FHWA Use of Data for Allocation Purposes
          Antonio E. Esteve, Consultant

2:15-2:30PM   Break

2:30-4:30PM   Session III: ***Alternative and Emerging Data Sources***
                     Hani Mahmassani, Moderator
                     Emerging Technologies for Highway Freight Monitoring
                     Catherine Lawson, SUNY Albany
                     ITDS: Future Directions
                     M.J. Fiocco, Department of Transportation
                     The Reebie Data Base: Measuring International Trade
                     Traffic
                     Paul Ciannavei, Reebie Associates
                     Measuring Highway Freight Traffic in Canada
                     Rob Tardif, Ontario Ministry of Transportation

FRIDAY, NOVEMBER 21, 2003

7:30-8:00AM   Continental Breakfast

8:00-8:30AM   Session III: ***Alternative and Emerging Data Sources***
                     (Continued)
                     Hani Mahmassani, Moderator
                     Freight Analysis Framework and Ton-Miles of
                     International Trade Traffic
                     Rolf Schmidt, Federal Highway Administration

8:30-10:15AM   Session IV: ***Statistical Quality***
                     Jim Lepkowski, Moderator
                     Error Profile of CFS
                     John Fowler, Bureau of the Census
                     Bill Davie, Jr., Bureau of the Census
                     Access to Export-Import Data for Allocation Formula Uses
                     Harvey Monk, Bureau of the Census
                     Alternative Data Source Quality and Availability
                     Daniel Melnick, Consultant

10:15-10:30AM   Break

10:30AM-12:00PM Open Discussion
                     Joel Horowitz

12:00-1:00PM   Lunch

Appendix
C

Workshop Presenters and Topics

William Bannister, Bureau of Transportation Statistics, "Origins of the 5115 Report"

Abstract: Bannister provided the history behind and motivation for the research that the Bureau of Transportation Statistics (BTS) had carried out on international trade traffic carried by highways that the panel reviewed. This history involved Section 5115 of the Transportation Equity Act of the 21st Century, and previous work in the late 1990s by Senators Moynihan and Chafee, and interactions between BTS, Department of Transportation, Department of Energy, the U.S. Census Bureau, and Oak Ridge National Laboratory.

Paul Ciannavei, Reebie Associates, "The Reebie Data Base: Measuring International Trade Traffic"

Abstract: Ciannavei provided an overview of the TRANSEARCH Freight Flow Database products, including estimates of domestic U.S. freight flows, U.S.-Canada cross-border data, and U.S.-Mexico cross-border data. Ciannavei described the sources for these data products and the methodologies used to produce them (including various kinds of adjustments). In addition, Ciannavei discussed highway network routings and inland movement patterns, and he provided a definition of international trade. Finally, he discussed empty truck movements and the distinction between nonfreight and freight.

Bill Davie, Jr., Bureau of the Census, "Error Profile of CFS"

Abstract: Davie provided a presentation on the major features of the Commodity Flow Survey, which is administered by the Census Bureau. This included the survey history, its objectives, industry coverage, sample design (three stages: establishments, weeks, and shipments), data items collected, editing and imputation methodologies, weighting, variance estimation, sample weights, response rates, and changes over time.

Antonio E. Esteve, Consultant, "FHWA Use of Data for Allocation Purposes"

Abstract: Federal-aid highway funds have been distributed to the states based on apportionment factors contained in highway legislation for some time. Esteve discussed the data systems used by the Federal Highway Administration to develop annual mileage and traffic apportionment factors as required by federal legislation in TEA-21, enacted in 1998. The programs described included the Highway Performance Monitoring System (HPMS), the Traffic Monitoring Guide (TMG), the American Association of Highway and Transportation Officials (AASHTO) Guidelines for Traffic Data Programs, and the Traffic Monitoring System (TMW) regulation. The factors include interstate lane-miles, interstate vehicle-miles, annual contribution to the highway account, lane-miles, vehicle-miles, diesel fuel used on highways, total lane-miles on principal arteries divided by a state's population, share of total cost to repair or replace deficient highway bridges, weighted nonattainment and maintenance area population, equal shares to each eligible state, nonhighway recreational fuel use during the preceding year, and urbanized area population. Esteve indicated how the various data systems might be used to estimate international trade carried by highways.

M.J. Fiocco, Department of Transportation, "ITDS: Future Directions"

Abstract: Fiocco presented an overview of the International Trade Data System (ITDS) and Automated Commercial Environment (ACE), which is intended to provide an Internet-based, integrated government-wide trade and transportation data capability. The hope is for this to become the government's front-end information technology system for all federal trade and border agencies and to provide a single interface with a harmonized federal data set for import, export, and in-transit information on cargoes, conveyances, and crews. The presentation included details on the vision for ITDS/ACE, details on both ACE and ITDS, how the system is being developed, primary uses and users, and the proposed operational environment. Fiocco concluded with the current status of ITDS and ACE, and the data implications of the development of this system.

Catherine Lawson, State University of New York at Albany, "Emerging Technologies for Highway Freight Monitoring"
Abstract: Lawson provided the current status of the several emerging technologies and data collection systems: (1) the Intelligent Transportation System (ITS), including the Archived Data User Service (ADUS); (2) work carried out by Battelle Transportation Research on use of global positioning system technology and memory cards; (3) the International Mobility and Trade Corridor (IMTC); (4) the Commercial Vehicle Information Systems and Network (CVISN); (5) AirTrak; and (6) the Idaho Department of Transportation Data Collection System. Lawson also discussed the possibility of the use of video imaging for data collection and SMART trucks.

Daniel Melnick, Consultant, "Alternative Data Source Quality and Availability"
Abstract: Melnick provided a description of the information available from four data systems: (1) Transborder Surface Freight, (2) PIERS, (3) railway weight bills, and (4) inland waterborne data from reports filed with the Corps of Engineers. Melnick offered the important distinction between information typically collected on trade and information typically collected on transportation, with the associated limits that trade data have for inferring transportation flows.

Paul Metaxatos, Universal Technical System, University of Illinois at Chicago, "Estimation and Accuracy of Origin-Destination Highway Freight Weight and Value Flows"
Abstract: Metaxatos described the gravity model in detail, including how the parameters are estimated, the sensitivity of the estimates to changes in the parameter estimates, needed data sources, model output—including confidence intervals, and results based on current data.

Harvey Monk, Bureau of the Census, "Access to Export-Import Data for Allocation Formula Uses"
Abstract: Monk discussed what import and export information could be shared with the Bureau of Transportation Statistics under the authority of Title 13 of the United States Code. Included was information on the types of data collected, the quality, and current access under Title 13.

Rolf Schmidt, Federal Highway Administration, "Freight Analysis Framework and Ton-Miles of International Trade Traffic"

Abstract: Bruce Lambert gave a presentation on the Freight Analysis Framework (FAF), which was created to understand the magnitude and geography of freight moving on the nation's transportation system, to develop a tool to evaluate emerging congestion, and to support reauthorization policy analysis. The FAF is a synthesis of diverse data from BTS, Army Corps, Reebie Truck, Rail Waybill Sample, etc., working across modes, to develop a better understanding of emerging logistics and various trade and transportation issues. The current products are a commodity database, a highway capacity database, and maps of freight activity for the nation, states, and selected metro areas.

Irwin Silberman, Bureau of Transportation Statistics, "Conceptual Underpinnings of Ton-Miles and Value-Miles"

Abstract: Silberman discussed the advantages and current uses of information on ton-miles. The strength is the combination in the statistic of two essential elements of transportation service, distance and quantity. Tonnage absent distance has the disadvantages of the possibility of double counting and the importance of the distance a load travels. Value-miles seem somewhat less useful as in discriminating since high value goods are most likely to be transported by air.

Frank Southworth, Senior R&D Staff, Oak Ridge National Laboratory, "Missing Pieces in Ton-Mile Estimation and ORNL Highway Network"

Abstract: This presentation added greater detail to the workings of the flow assignment model described by Hwang. The primary step involves selection of the most likely origin-destination paths through the ORNL multimodal network, using either a least impedance path model or a logit-based weighted distance average. The presentation also included the sources of uncertainty of the distance calculations.

Rob Tardif, Ontario Ministry of Transportation, "Measuring Highway Freight Traffic in Canada"

Abstract: This presentation discussed a variety of datasets that are used to help measure highway freight traffic in Canada. They include the Canadian Vehicle Survey and trade data.

Appendix
D

Biographical Sketches of
Panel Members and Staff

JOEL L. HOROWITZ *(Chair)* is the Charles E. and Emma H. Morrison professor of economics at Northwestern University. He has had previous positions at the University of Iowa and the U.S. Environmental Protection Agency. His primary area of research is in theoretical and applied econometrics, with particular concentrations in semiparametric estimation, bootstrap methods, discrete choice analysis, and estimation and inference with incomplete data. He received a B.S. in physics from Stanford University and a Ph.D. in physics from Cornell University. He is coeditor of *Econometrica* and the former coeditor of *Econometric Theory*. He is an elected fellow of the Econometric Society, a winner of the Richard Stone Prize in applied econometrics, and a recipient of the Alexander von Humboldt Award for senior U.S. scientists.

JAMES M. LEPKOWSKI is a research professor at the Institute for Social Research and an associate professor of biostatistics, both at the University of Michigan. He is also a research professor in the Joint Program in Survey Methodology at the University of Maryland. He received a B.S. in mathematics from Illinois State University and a Ph.D. in biostatistics from the University of Michigan. He conducts survey methodology research, including the design and analysis of area probability and telephone samples, compensating for missing data, and telephone sampling methods. He is a fellow of the American Statistical Association and was elected to membership in the International Statistical Institute. He is a member of the National

Academies Committee to Review the Bureau of Transportation Statistics' Survey Programs.

HANI MAHMASSANI is the Charles Irish senior chair in engineering and director of the Maryland Transportation Initiative, both at the University of Maryland. He specializes in transportation systems modeling and the application of advanced operations research and econometric techniques to the analysis, design, optimization, and operation of transportation systems. He received an M.S. in transportation engineering from Purdue University and a Ph.D. in transportation systems from the Massachusetts Institute of Technology. He chairs the Traffic Flow Theory and Characteristics Committee of the Transportation Research Board of the National Academies, and he is associate editor of *Transportation Science.*

AMELIA REGAN is an associate professor in the Department of Civil and Environmental Engineering and Graduate School of Management at the University of California, Irvine. Her research has been in applications of information technologies and optimization techniques to freight and fleet management, transportation logistics, intermodal operations, and commercial vehicle operator and firm behavior. She received a B.S. in systems engineering from the University of Pennsylvania, an M.S. in applied mathematics from Johns Hopkins University, an M.S.E. in civil engineering from the University of Texas, and a Ph.D. in transportation systems engineering from the University of Texas.

BRUCE DAVID SPENCER is a professor in the Department of Statistics and a faculty fellow at the Institute for Policy Research, both at Northwestern University. Prior to joining the staff of Northwestern University, he was senior research statistician for the National Opinion Research Corporation of the University of Chicago. He received a B.S. degree in biometry from Cornell University, an M.S. in statistics from Florida State University, and a Ph.D. in statistics from Yale University. He is a fellow of the American Statistical Association. He has participated in evaluations of major statistical programs—including population estimates by the Census Bureau, population forecasts by the Social Security Administration, test score statistics by the Department of Education—and drug abuse estimates by state and local agencies. He has also conducted research into the effects of data error on the allocations of public funding and representation.

RONALD TWEEDIE retired in 2001 from the New York State Department of Transportation, where his positions included director of the Planning Bureau and director of the Data Services Bureau. He was responsible for the development and direction of the state's comprehensive transportation planning program and for the coordination of program activities with other state agencies, metropolitan planning organizations, the federal government, and local jurisdictions. He also directed the activities of staff providing transportation data and analysis services essential to developing capital projects, setting priorities, and allocating funds in accordance with state procedures. He chairs the National Academies' Committee on Statewide Transportation Data and Information Systems and is a member of the Committee to Review the Bureau of Transportation Statistics' Survey Programs. He holds a B.S. in civil engineering from the Massachusetts Institute of Technology, and an M.P.A. from the State University of New York at Albany.

TOM PLEWES (*Co-Study Director*) is a senior program officer for the Committee on National Statistics. In addition to the Committee for the Review of Research and Development Statistics at the National Science Foundation, he is directing studies of international trade traffic statistics and working on National Research Council (NRC) initiatives with the U.S. General Accounting Office on key national indicators of performance. Prior to joining the NRC staff, he was associate commissioner for employment and unemployment statistics of the Bureau of Labor Statistics. He is a fellow of the American Statistical Association and was a member of the Federal Committee on Statistical Methodology. He has a B.A. degree from Hope College and an M.A. degree from the George Washington University.

MICHAEL COHEN (*Co-Study Director*) is a senior program officer for the Committee on National Statistics, currently also serving as study director of the Panel on Correlation Bias and Coverage Measurement in the 2010 Census and co-study director of the Panel on the Functionality and Usability of Data from the American Community Survey. He was co-study director of the Panel on Research on Future Census Methods and assisted the Panel to Review the 2000 Census. He has also served on the staff of the Panel on Operational Test Design and Evaluation of the Interim Armored Vehicle (Stryker), the Panel on Estimates of Poverty for Small Geographic Areas, and the Panel on Statistical Methods for Testing and Evaluating

Defense Systems. Formerly, he was a mathematical statistician at the Energy Information Administration, an assistant professor in the School of Public Affairs at the University of Maryland, and a visiting lecturer in statistics at Princeton University. His general area of research is the use of statistics in public policy, with particular interest in census undercount, model validation, and robust estimation. He is a fellow of the American Statistical Association. He received a B.S. degree in mathematics from the University of Michigan and M.S. and Ph.D. degrees in statistics from Stanford University.